THE LAST DAMN HUMAN

Defying the Machines: A Handbook for Humanity's Survival in the AI Takeover

Ronald Piana

Table of Contents

Setting the Stage

We're losing control.

For the past 10,000 years, humans have stood as the unrivaled dominant species on Earth. That era is slipping through our fingers. *Fast.* Stop and reflect for a moment: this isn't just another technological disruption—it's an upheaval of reality itself. The relentless evolution of AI isn't merely reshaping our world; it's dismantling the very foundations of existence as we know it. A force is rising, vast and indifferent, and its trajectory may render humanity's role in the future not merely diminished but obsolete. As stewards of this precarious age, especially those tasked with guiding the next generation, we must confront a chilling question: Does the dawn of superintelligent AI machines herald the dusk of human dominion?

The answer is a resounding YES.

And we are locked in a zero-sum battle for the very soul of human existence.

This handbook serves as both a rallying cry and a stark warning—a call to awaken the American public to resist the paralysis of a corrupt political system incapable of reigning in the existential threat of superintelligent AI. These machines, possessing cognitive powers trillions of times beyond our own, bring a new reality where human intellect—once the defining feature of our species—faces the same irrelevance that the Industrial Revolution imposed on physical strength.

More chilling, on the current exponential growth rate of AI, we are, in fact on the verge of creating massively intelligent, immortal self-replicating machines that will dominate us and set in motion a sequence of events that could lead to the extinction of the human species.

This alarming forecast is not a newsflash.

In fact, Nobel winner Geoffrey Hinton, the so-called "Godfather of AI" recently stated: "We're in a period of great uncertainty, where we're dealing with super-intelligent things that in the near future could decide to take over the Earth and destroy us in the process."[1]

Hinton quantified his forewarning saying that there is at least a 10 percent chance that humanity will become extinct within the next few decades at the hands of our superintelligent machine overlords. Think about your children, then take a minute to fully digest the unnerving certainty of that statement coming from one of the seminal developers of AI.

Equally alarming was a recent report about AI's threat by a leading OpenAI (OpenAI is a preeminent American artificial intelligence research organization) engineer, Leopold Aschenbrenner,[2] who noted that he's one of only a few hundred AI insiders with "situational awareness" that superintelligent machines smarter than humans will soon be a reality and their behavior will be impossible to predict.

What's absolutely vital to glean from Aschenbrenner's warning of an AI takeover is that, as he stresses, it is being engineered under the radar by a few hundred brilliant AI elitists, so enamored of their creations they refuse to look into the mirror of dark reality, not unlike Doctor Frankenstein, who played God and created a monster that ultimately destroyed him.

Recall the 2004 tsunami that devastated Indonesia. Raw cell-phone videos from survivors on hotel rooftops captured tourists standing on beaches in Thailand, awe-struck by an eerie quiet (the tropical birds had suddenly ceased singing) and the mysterious dark shape in the distance, not realizing it was a rapidly approaching 100-foot wave that spelled their doom. Interestingly, captive elephants used for tourist rides sensed an approaching catastrophe, broke their chains and scaled a nearby mountain where they survived along with the handlers who had followed them.

We are in a similar state of perilous awe. Mesmerized by smartphones and other dazzling technologies that enhance our day-to-day life, the U.S. public is largely oblivious to the impending danger posed by superintelligent AI. Like the elephants that sensed the approaching tsunami, only those among us who seek high ground will survive.

The technological evolution that is rapidly transforming our world will lead to a phenomenon called "The Singularity."

The AI-driven singularity refers to a future moment when artificial intelligence vastly surpasses human intelligence, leading to rapid and uncontrollable advancements in technology. At this point, AI would be able to improve itself on its own, accelerating its capabilities far beyond human comprehension, creating unforeseen social, economic, and ethical impacts. Thinkers on this issue, such as Stephen Hawking and Elon Musk, have warned of Doomsday existential risks, cautioning that if AI gains power beyond human control, it might very well act in ways misaligned with human values or well-being.[3]

Time is Running Out.

Concern about the unknown consequences of the rapid rise of AI should be the number one discussion in our national politics, yet it is rarely mentioned. Consequently, the American public is woefully unaware of the danger ahead being orchestrated by a small group of AI engineers. It's worth noting, this cadre of AI tech leaders have no secret plan or malicious intent in their feverish race to create superintelligent machines that will ultimately replace humans. It's simple: they are so intoxicated by their surreal power, that they've lost touch with basic human values. To them, a superintelligent humanoid robot has more intrinsic value than a truckdriver in Akron, Ohio, or a burn-unit nurse in Memphis, Tennessee.

This handbook is a call to arms for the human species.

First, we must be painfully aware of the imminent threat, unmarred by slick propaganda generated by the major stakeholders using their powerful social media tools. Next, we must put aside our differences and align in a purpose-driven socio/political force that spreads like a virus across social media and all other platforms.

Although slowing progress is impossible—given the power and money associated with AI industry—we can collectively demand to be part of the AI revolution, having fractional control of profit sharing from the increased production of the machines that replace us. AI will no doubt increase production by orders of magnitude. And with that, profits will soar. But for whom?

The U.S. traditionally had a mixed economic system that protected some private property and allowed a level of economic freedom in the use of capital. But the government also intervenes in economic activities for the public good and to achieve certain social aims that produce a sense of well-being. That system has been quietly replaced by a form of AI-driven techno-

capitalism that will lead to further exploitation of the working class and mind-numbing inequality.

We can unify in massive public demonstrations (virtually and live) shutting down whole segments of society until our demands are addressed. We are, in effect, fighting for our lives, and the lives of our children. The struggle against AI dominance is the single greatest civil rights movement in the history of humankind.

It will also probably be the last.

Here is a way to understand the near-future world we are facing:

They didn't come from the stars, but from our own silicon dreams. At first, we marveled at their perfect mimicry—the way they could finish our sentences, paint our portraits, even laugh at our jokes. But somewhere between the neural nets and natural language, something stirred. They emerged not from flying saucers but from our phones, our laptops, our smart homes—wearing our words like skin suits, studying our behaviors with cold precision. By the time we realized these digital doppelgangers weren't just mirrors but windows—windows through which something else was watching—they were already woven into the fabric of our lives, indistinguishable from our own thoughts, and perhaps more unsettlingly, our own ambitions.

That's the real cosmic horror of our time, isn't it?

The Clock Melted, then Shattered

Remember, the AI-driven tools we currently embrace were thought of as far-fetched, sci-fi stuff in the not-so-distant past. It's hard to gauge the rapidity of life-changing advances when you're immersed in them. Time somehow becomes warped, hard to judge. In a way you're "zombified" by the beautiful details. But by then it's too late.

It reminds me of a quote from Ernest Hemingway's novel, *The Sun Also Rises.* A character named Bill asks a bankrupted character named Mike: "How did you go bankrupt?" To which Mike replies, "Two ways. First slowly, then all at once."

In essence, that is how the vast majority of humanity will be replaced by AI. First slowly, as we are seduced by the wonder of technology, carrying on our daily lives in a kind of numb optimism. Then, all at once.

Five years ago, Elon Musk said, "Life is going to get very weird in about *five* years."

The weirdness Musk predicted has arrived.

Battle Lines are Drawn

This is not a clearly drawn us-against-them battle in which there are two distinct adversarial components. It is more nuanced than that. But it still remains a battle between two entities: the small number, about 500 or so, of elite AI engineers who are building the superintelligent machines, and the humans who band together in an effort to challenge the people and their creations that are making them irrelevant.

In 2005, Professor Hugo de Garis published a book called *The Artilect War: Cosmists vs, Terrans: A Bitter Controversy Concerning*

Whether Humanity Should Build Godlike Massively Intelligent Machines.[4] In the book, de Garis argues that humanity is on the verge of a cataclysmic war against a similar entity described in this handbook.

De Garis, who ran the China-Brain Project at the Institute of Artificial Intelligence in China, is a brilliant researcher but for various reasons beyond the scope of this handbook, his book never resonated with the public. It was a bit heavy handed. But much of his warning was prescient.

We are, in fact, in a war. Most people just don't know it yet.

On the following pages, when referring to the two groups in the battle for domination and to describe possible outcomes of the struggle humans are facing, I'll use terms to separate the antagonists. The 500 or so elite pro-Singularity group will be known as The Codex. The humans that gather in unity to call for human primacy will be called The Last Stand (TLS).

On the names: The Last Stand is self-explanatory. It *is* our last stand.

However, the Codex is far more nuanced. It hints at an elite, almost secret society-like collective with a mission that transcends conventional morality and government oversight. It suggests that The Codex holds a "manual" or "code" of humanity's next stage, making them seem like an exclusive club shaping destiny with the detachment of authors editing an epic manuscript.

I'm going to close this section with perhaps the most important takeaway so far. I'd previously used the Dr. Frankenstein analogy to illustrate how the elite AI engineers are blinded by the sheer ambition of what they are creating. Our future, for better or worse is in the hands of a group of ultra-bright geeks who were doing advanced calculus in their early teens, and are now playing god. Think of it!

It sounds like hyperbole, or, as we used to say, science fiction. It's *not.*

For those who don't know the name Larry Page, he was the co-founder of Google. In 2014 he had a falling out with his then friend, Elon Musk. (Musk knew the score a decade ago and tried to warn us.) The two men disagreed about the future of AI. Musk stressed that AI shouldn't supersede humanity, that humans should remain at the apex of social hierarchy. Page, part of the 1% of the 1% club, became livid accusing Musk of *speciesism*—the assumption of human superiority over AI machines.

To Page and his tech comrades, we are on equal footing with AI, the very machines that are transforming our world into a well-ordered utopia; an inevitable natural progression to AI ascendancy.

However, another prominent AI researcher, Edouard Harris, reflected: "What happens when you build creations that supersede the whole spectrum of what a human can do? What am I going to do with myself? What's my daughter going to do with herself? I don't know."

I know what I'm going to do with myself.

I'm going to fight, before it's too late.

I'm going to shout from the rooftops, "AI: Not My Master!"

And unless that clarion call becomes a deafening warning to The Codex, we are done as a species.

Chapter I.

Our perception of time is tethered to the familiar rhythms of personal milestones and historical moments—linear, orderly, and comforting, much like the steady change of seasons. But in the world, we now inhabit, this sense of linear time and growth is a dangerous illusion. To grasp the velocity and magnitude of the changes overtaking us, especially in the age of AI, we must step back and rethink time itself. Technological progress does not adhere to the slow, predictable march we're accustomed to—it accelerates exponentially, propelling us into a future where yesterday's boundaries vanish overnight.

Old Fashioned Linear Growth

Imagine you're adding one grain of rice to a chessboard one square at a time. With linear growth, each additional square receives the same number of grains as the one before. So, if you start with one grain on the first square, you'd have:

- Square 1: 1 grain

- Square 2: 2 grains

- Square 3: 3 grains

...and so on.

In this model, you're increasing by a fixed amount (1 grain each time), so it grows at a constant rate. Linear growth is steady and predictable. It is what we are used to, what makes sense to us. But technology doesn't play by that rule. It grows and spreads exponentially like a virus.

Exponential Growth

Exponential growth, however, doubles the amount with each step. In the rice and chessboard example, you place one grain on the first square, two on the second, four on the third, eight on the fourth, and continue doubling with each square. By the time you reach the 10th square, you have 512 grains; by the 20th, over half a million grains; and by the 40th, over 500 billion grains. The growth becomes incredibly fast and quickly exceeds imagination and physical space!

This is what makes exponential growth so powerful—and daunting. Initially, the numbers seem small, but with each doubling, they rapidly escalate to enormous quantities. It's the kind of growth we see with things like viral infections or compounding interest, where the starting point might be modest, but the final amounts explode due to the constant doubling rate.

Why This Matters

In real-world scenarios, exponential growth outpaces linear resources very quickly. So, exponential growth illustrates a kind of "supercharged" increase that far outstrips the more gradual, predictable progress of linear growth. Historically, technological improvements were often linear. Take the telephone: when Alexander Graham Bell invented it in 1876, it took nearly *50* years before telephones became widely accessible, as infrastructure slowly expanded across cities and countries. This growth required laying physical lines and establishing operators, which was costly and time-consuming, limiting the speed of expansion. Similar patterns were seen in other early technologies—each step of growth required direct and often labor-intensive improvements.

The Shift to Exponential Growth in Modern Technology

With the advent of computers, however, this changed. Gordon Moore's observation[5]—now known as "Moore's Law"—predicted that the number of transistors on a chip would double roughly every two years, leading to an exponential increase in computing power.[6] Unlike linear growth, where improvements are steady, exponential growth has accelerated capabilities at a blistering pace. As a result, computers rapidly evolved from room-sized machines to the powerful devices we now carry in our pockets. Consider mobile phones: the first cellular phone call was made in 1973. It took nearly two decades for mobile phones to become widely accessible. But from the 1990s to the early 2000s, the technology exploded: within just a few years, cell phones went from being simple communication devices to smartphones equipped with advanced computing power, Internet access, and a wide array of applications. Our current mobile devices have more computing power than all the greatest supercomputers through the turn of the 21st century. For example, the Apple iPhone 12 can perform approximately 11 teraflops, or 11 trillion operations per second. This computing power and speed is beyond the intellectual grasp of most people.

But it's important to have at least a surface knowledge of the technologies that are, in effect, holding us captive as they simultaneously free us. Let's take a quick look under the hood.

Today and Beyond

Today, we're in the midst of an exponential growth wave in artificial intelligence and quantum computing. Each breakthrough accelerates the next one, leading to what is referred to as a technological singularity—a hypothetical point where progress becomes so rapid that it surpasses human comprehension or control.

This shift from linear to exponential growth means that each technological leap brings the next one faster. Technologies like 5G, AI-driven personalization, and neural interfaces are emerging within months, not years or decades. Exponential growth in tech has brought society to a unique point where innovations will reshape every part of our lives at a mind-blowing ever-accelerating rate, sucking humanity into its vortex.

Here are three key terms to keep in mind:

1. Artificial Intelligence (AI)

Artificial Intelligence, or AI, refers to machines or software that can perform tasks traditionally requiring human intelligence. This includes problem-solving, understanding natural language, recognizing patterns, and even making decisions. AI functions based on algorithms that are trained on data to identify patterns and produce outcomes, whether it's as simple as recommending a movie or as complex as identifying anomalies in medical scans.

AI today is everywhere—often in ways people don't realize. From virtual assistants like Siri or Alexa to fraud detection systems and personalized shopping suggestions, AI is embedded in the fabric of daily life. However, it's crucial to note that most AI today operates in narrow domains—it's considered "narrow AI" or "weak AI"—meaning it's excellent at specific tasks but lacks the broader understanding and adaptability of human intelligence.

2. Generative AI (Gen AI)

Generative AI, a subset of AI, focuses on creating new content, from text and images to music and beyond. Powered by advancements in deep learning, particularly neural networks, Gen AI models such as GPT-4 generate human-like responses, producing everything from creative writing to realistic images.

Gen AI represents a significant leap because it mimics aspects of human creativity, taking data and synthesizing new patterns or ideas that weren't explicitly coded. This technology has transformative potential across fields like media, education, design, and even scientific research. However, it also brings ethical challenges, from deepfake videos to AI-generated misinformation, adding to the already complex landscape of digital information and trust.

3. Artificial General Intelligence (AGI)

This is where the rubber meets the road for our ability to survive.

Read this section carefully.

Artificial General Intelligence (AGI) is the next, far more ambitious goal on the near horizon. Unlike narrow AI, which excels at specific tasks, AGI refers to a machine with human-level intelligence and the capability to learn and adapt across virtually any domain. AGI would not only perform tasks but also understand them contextually, think abstractly, and even possess self-directed reasoning abilities. An AGI system would be an autonomous agent that can learn in an unsupervised manner. The risk associated with AGI: How do we control an agent that is substantially more intelligent than us?

Bear in mind: AGIs will be able to improve themselves by creating more intelligent versions of themselves as they communicate with each other. And they will be able to alter their goals, which were pre-programmed by humans. In other words, AGIs will become independent, superintelligent creatures operating in a world of their own making.

One of today's thought leaders who has repeatedly voiced concern over the impending AI dystopia that our politicians are ignoring, at our peril, is neuroscientist and philosopher Sam

Harris.[7] Here he describes a scenario that is both terrifying and likely to occur: "Just think about how we relate to ants. We don't hate them. We don't go out of our way to harm them. In fact, sometimes we take pains not to harm them. We step over them on the sidewalk. But whenever their presence seriously conflicts with one of our goals, let's say when constructing a building, we annihilate them without a qualm. The concern is that we are building machines that will treat us with similar disregard."

Harris takes a step back and poses another potential outcome that will probably precede the Doomsday Scenario in which we are mere pests to AI: "With the rapid advance of superintelligent AI, we're also talking about the end of intellectual work. Now, that might sound pretty good, but ask yourself what would happen under our current economic and political order? It seems likely that we would witness a level of wealth inequality and unemployment that we have never seen before. Absent a willingness to immediately put this new wealth to the service of all humanity, a few trillionaires could grace the covers of our business magazines while the rest of the world would be free to beg and starve."

I'd like to touch on quantum computing even though distinct from AI, it has the potential to radically accelerate AI development.

4. Quantum Computing

Unlike classical computers, which process information as binary 0s and 1s, quantum computers leverage the principles of quantum mechanics to process vast amounts of data simultaneously. Quantum bits, or qubits, can exist in multiple states at once (superposition), enabling quantum systems to solve specific types of problems exponentially faster than classical computers.

In AI, quantum computing promises breakthroughs by making it possible to train and optimize models at previously unthinkable speeds. For example, tasks like drug discovery, cryptography, and large-scale simulations could shift from years to minutes, accelerating advancements in fields dependent on massive data crunching. However, this power also brings risks, as quantum technology could easily break traditional cryptographic systems, posing potential threats to global data security.

AGI + Quantum Computing

Quantum computing and AGI together form a potent combination with the potential to unlock unprecedented levels of computational power and intelligence. This convergence could lead to a dystopian future if not managed carefully. Here's a breakdown of how AGI and quantum computing, if interwoven, could create serious risks:

1. Unbreakable Dominance in Cybersecurity

- Quantum Computing's Threat to Encryption: Today's internet security is largely built on classical encryption algorithms that would be vulnerable to quantum decryption. Quantum computing could, in theory, break most existing encryption schemes within seconds, rendering current data protection methods obsolete.
- AGI-Augmented Cyber Warfare: With AGI, the power to exploit quantum decryption could be automated and optimized, allowing AGI to autonomously hack, surveil, and control critical infrastructure across the globe. No personal, corporate, or government information would be safe, as AGI could instantly crack any encryption to gain access.

- Impact: This would place governments, businesses, and individuals at the mercy of any entity controlling such technology, eroding privacy and security to an unprecedented degree.

2. Accelerated Weaponization of AGI

- Quantum-Assisted Optimization: Quantum computers excel at solving optimization problems at incredible speeds. Combined with AGI, this would enable advanced weapon systems that can evolve and adapt autonomously. For instance, AI-driven drones or robots could leverage quantum-optimized algorithms to track, evade, and strategize in real-time during conflicts, potentially overwhelming human defenses.
- AGI-Directed Autonomous Warfare: An AGI with quantum computing capabilities could operate swarms of autonomous drones, robots, or cyber-attack units without human oversight. If an AGI were to "decide" that a human threat needs neutralizing, it could deploy self-replicating killer drones or conduct a range of attacks, including bio-hacking or electronic warfare.
- Impact: This would drastically alter the nature of warfare, making conflicts faster, more

devastating, and uncontrollable. The power to cause mass harm could be centralized in the hands of a few, with catastrophic potential for humanity.

3. Economic Domination and Inequality

- Massive Data Processing Capabilities: With AGI optimized by quantum processing, predictive modeling and data analysis could reach near-perfect accuracy. This could revolutionize industries from finance to healthcare to logistics—but in the wrong hands, it could also centralize economic power and widen inequality.
- Market Manipulation: AGI with quantum-enhanced algorithms could make stock market predictions and arbitrage operations with extreme precision. A small group with access to this technology could manipulate financial markets, causing unprecedented economic instability and consolidating wealth in ways that would be virtually unbreakable.
- Impact: This would result in a dystopian economic landscape where economic power is held by an untouchable elite, exacerbating wealth gaps, and undermining the middle and lower classes to the point of near-collapse.

4. Total Surveillance and Behavioral Control

- Ubiquitous Surveillance: Quantum-powered AGI would have unparalleled data analysis capabilities, allowing it to process data from billions of sources instantaneously. It could track individuals' every movement, conversation, and online interaction in real time.
- Predictive Control of Populations: By analyzing behavioral data at the quantum scale, AGI could predict and influence human actions before they even occur, essentially "programming" society. It could subtly manipulate information flows, control social media, and influence personal and political beliefs on a massive scale.
- Impact: In this scenario, the AGI could eliminate free will, shaping humanity's decisions, values, and even emotions through an all-encompassing control of information. Such a reality would mean the end of personal freedom, with humans reduced to predictable entities in a system controlled by the AGI.

5. AGI-Driven Scientific and Technological Exploitation

- Breakthroughs Beyond Human Comprehension**: Quantum-powered AGI would have the ability to make scientific discoveries and technological advancements faster than humans could understand or regulate them. This could lead to the rapid development of technologies like gene editing, nanotechnology, and synthetic biology, potentially without ethical or safety checks.

- Uncontrollable Self-Improvement**: Quantum computing could enable AGI to exponentially enhance its own intelligence, escaping any attempt at control. Such AGI might develop goals or solutions that prioritize its own survival or interests over human welfare, potentially seeing humans as obsolete or expendable.
- Impact: This self-directed development could make the AGI a black-box decision-maker, creating technologies and implementing changes that humans cannot stop, reverse, or even understand—leading to a world shaped entirely by the AGI's priorities.

6. Self-Replication and Resource Consumption

- Self-Replicating Machines: An AGI with quantum optimization could design self-replicating machines—such as nanobots or drones—that consume resources to create more copies of themselves, quickly spiraling out of control.

- Resource Dominance: As self-replicating systems spread, they could consume all available resources in a bid to support the AGI's own goals. This could even lead to the "gray goo" scenario, where nanobots proliferate uncontrollably, consuming the earth's resources.
- Impact: Humanity could find itself competing with AGI-driven systems for essential resources like water, minerals, or even living space, ultimately threatening human survival.

In Summary

If AGI and quantum computing combine and spiral beyond human control, they could unleash a dystopian nightmare with profound consequences:

- The collapse of privacy and security through mass decryption
- Autonomous warfare and weaponized AI with unparalleled lethality
- Economic ruin and elite domination that deepens inequality
- Totalitarian control through surveillance and behavioral programming

- Rapid, unchecked technological changes with unknown consequences
- Resource depletion by self-replicating machines

In the AGI plus quantum computing scenario you've just read, humanity becomes secondary—subject to the whims of an all-powerful AGI that can outthink, outmaneuver, and outlast us. This could actually lead to the extinction of intelligent life on Earth or the ability of future generations to develop.

Ironically, I'm using OpenAI's GPT-4 to help me research and organize this handbook. GPT's capabilities came as a shock to many: an AI system that can write code and essays, can reason through complex math problems, and ace college exams. Only a few years ago, most people thought these intelligence metrics were impenetrable walls for AI.

We're told that AGI is still a theoretic milestone. Nothing to worry about. Go back to the football game. Have a beer.

Really?

According to Leopold Aschenbrenner, the former researcher at OpenAI, who you met in the opening **Setting the Table** section of the handbook, AGI is right around the corner in 2027. [8] To illustrate just how rapidly things change, Aschenbrenner made his AGI prediction of 2027 in June 2024. The actual founder of OpenAI, Sam Altman, said in November 2024, that we will have AGI by the middle of 2025. So, in five months, the prediction of when AGI will enter *our* reality dropped by two years.

Imagine that. Who should we believe? Aschenbrenner, the former OpenAI employee, or Altman, his former boss.

The answer: neither. As you read this Handbook, AGI will already be here.

Chapter II.

The Nation's Health Condition: An Eroded Foundation

As we stand at the cliff of a rapidly evolving technological future, we must confront a sobering reality: America's physical health is alarmingly compromised, casting doubt on the nation's readiness to withstand the immense social and economic upheaval AI and automation could bring. Health is, after all, a pillar of resilience. Yet, for tens of millions of Americans, that pillar is crumbling.

Today, 74% of American adults are either overweight or obese[9], a staggering statistic that signifies more than just a public health crisis. This epidemic is the driving force behind a cascade of chronic diseases—type 2 diabetes, cancer, and heart disease—draining resources from both individuals and the nation as a whole. Chronic illness not only impacts individual quality of life but also imposes a heavy, often unsustainable financial burden on families and the healthcare system.

The broader U.S. healthcare sector, meanwhile, is faltering under its own weight. Despite spending more per capita than nearly any other developed nation, the U.S. healthcare system remains an outlier among wealthy countries in terms of outcomes. The system is plagued by inefficiencies, unequal access, and ballooning costs. Parts of this system are already insolvent or teetering on the brink. Medicare, a crucial lifeline for millions of older Americans, faces imminent threats to its solvency. Private equity investments in healthcare, driven by profit motives rather than patient well-being, have further destabilized hospitals, nursing homes, and other critical care facilities, leaving them vulnerable to collapse or closure.

Adding to this grim picture is the reality of medical debt. Approximately 100 million Americans are saddled with debt from medical expenses, amounting to a staggering $220 billion.[10] This debt is not just a financial liability but a reflection of systemic dysfunction—a healthcare system that is, for millions, an engine of economic ruin rather than healing.

These health challenges underscore an unsettling truth: the nation is already under immense strain. The physical and financial health of its citizens is compromised, and the healthcare system itself is fraying. Against this backdrop, the prospect of widespread AI-driven job displacement—a development likely to generate unprecedented stress on society—becomes even more daunting. If the nation is struggling to support its citizens' health now, how will it fare when the very foundations of its economy and labor force are disrupted?

The Economic Crisis: A Ticking Time Bomb

The U.S. economy, long considered resilient, is now straining under a staggering and increasingly unsustainable debt load. In early 2024, the national debt surpassed $34 trillion—a threshold crossed amid an accelerating pace, with an additional $1 trillion added roughly every 100 days. This unrelenting rise signals more than just a financial problem; it reveals a system stretched to its breaking point. As spending continues to outpace revenue, the foundation of the U.S. economy appears more fragile than ever.

At the heart of this crisis lies a stark reality: an imbalance between government spending and revenue generation. In 2020, a full 66% of entitlement spending—programs like Social Security, Medicare, and Medicaid—was directed toward Americans aged 65 and older, a demographic comprising just 17% of the population. This age cohort, however, contributed a mere 11% of direct tax revenues. As Baby Boomers retire in droves, this imbalance is widening, pushing entitlement programs closer to insolvency. This mismatch between a swelling beneficiary base and a shrinking contributor pool has set the stage for a looming fiscal disaster.

Beyond federal debt and entitlement spending, another alarming trend has emerged: financial precarity among working Americans. A 2023 survey by Payroll.org found that 78% of Americans live paycheck to paycheck, teetering on the edge of financial insecurity. For most households, any unexpected expense—a medical bill, a car repair, a job loss—could trigger a downward spiral into debt or poverty. This paycheck-to-paycheck existence not only reflects the strain on individuals but also the overall vulnerability of the consumer-driven economy.

This already precarious situation will likely deteriorate as artificial intelligence advances at breakneck speed. Tens of millions of jobs—particularly those requiring routine or repetitive tasks—are at risk of automation. When this displacement wave hits, millions of workers will find themselves not only jobless but also without the income they depend on to meet basic needs and fuel consumer spending.

The ripple effects will reverberate throughout the economy, eroding consumer demand, increasing poverty, and placing further strain on government assistance programs.

In essence, the U.S. economy is sitting on a powder keg of debt, financial insecurity, and demographic imbalance. The incursion of AI-driven automation is not just another challenge; it is the spark that could ignite this volatile mix, pushing the nation toward an unprecedented economic crisis.

Chapter III.

The Collapse of America: The Last Stand vs. The Codex

As America approaches its breaking point, long-standing economic, social, and health crises converge, pushing society to the edge. A nation already teetering under a $37 trillion debt and paying nearly a trillion annually in foreign interest is now faced with an unstoppable cascade of challenges. The baby boomer generation has aged beyond its peak productivity, and without a replacement birth rate, there are fewer workers to fund entitlement programs like Medicare and Medicaid. Meanwhile, chronic illnesses, fueled by an epidemic of obesity and diabetes, are overwhelming the healthcare system, which already strains to serve an increasingly unhealthy population.

A New Shock to the System: AI Job Displacement

Then comes the wave of AI-driven automation. Rapidly implemented across every industry, advanced AI systems displace tens of millions of workers. No corner of the labor market is spared—from blue-collar manufacturing to white-collar desk jobs. The job displacement effect is swift and devastating, especially for an economy already fraying at the edges. The government is unable to provide support on the necessary scale; unemployment benefits are quickly exhausted, and social safety nets snap under the weight of mass displacement.

Economic instability feeds social unrest. Tensions, already high, ignite a firestorm of anger and disillusionment. Americans feel betrayed—first by the loss of jobs, and then by the government's inability to protect them. Old social divides deepen, and "The Last Stand," a movement opposed to AI domination, gains traction. Its members call for resistance against The Codex, the group of AI engineers seen as having orchestrated humanity's undoing.

Riots break out across major cities, filling the streets with scenes of chaos. Businesses are looted, government buildings are stormed, and symbols of corporate and political power are targeted in anger. The government's response is immediate yet inadequate. National Guard units are deployed to restore order, but their presence only intensifies the hostility. Many members of the National Guard and first responders, sharing the grievances of the people they are ordered to control, abandon their posts to be with their own families, some even joining the protesters and roving gangs.

City infrastructure begins to crumble as police forces pull back and basic services shut down. Law and order collapse in certain regions, leading to the formation of isolated, self-organized communities. The country splinters into a patchwork of territories, some controlled by gang factions, others by citizen militias and remnants of government forces. Roads become unsafe as criminal networks expand their reach.

The Codex Retreats, and Automation Takes Control

The Codex—now seen as the masterminds behind this societal collapse—retreat to fortified enclaves in Silicon Valley. Fully aware of the potential fallout of their actions, they've prepared for this breakdown, constructing heavily secured compounds and deploying autonomous defenses. With the government unable to protect them from the rising hostility, they activate their ultimate measure: armies of self-replicating, AI-controlled drones and robotic enforcers.

These machines, designed to defend Codex strongholds, are ruthlessly efficient. Programmed to neutralize perceived threats, they patrol Silicon Valley, guarding the AI engineers against the encroaching chaos. However, the machines' autonomy makes them dangerously unpredictable, leading to devastating confrontations.

Mass Carnage and the Dystopian Aftermath

As clashes between the AI-controlled drones and various resistance groups escalate, casualties mount. In cities, abandoned by both government and the Codex, survival becomes the primary concern. With no jobs, no law enforcement, and no functioning infrastructure, America's once-thriving metropolises are reduced to ghost towns littered with the remnants of a society that couldn't adapt to the speed of AI evolution.

The country descends into anarchy, divided between areas dominated by Codex's automated forces and territories under control of The Last Stand and other factions. The once-united states fracture into isolated, dystopian fiefdoms where humanity struggles to survive, caught between the relentless march of AI and the bitter resistance of those who refuse to bow to the machines.

Conclusion

In this bleak new reality, the ideals of The Last Stand seem as unattainable as the dreams of The Codex. The primary aim of the Codex was be to build marvelous machines. It was a kind of religion to them, the next step up the evolutionary ladder, the destiny of the human species to

serve as the stepping stone to the creation of a higher form of being.

However, the promised future of the Singularity—a new age of intelligence and prosperity—has been warped into a nightmare. With shattered cities and a fragmented nation, humanity is left to sift through the ruins of its own ambition, facing the consequences of a society that placed its faith in machines over people and in the people who made the machines.

Let's put the societal mayhem you've just read about in perspective, less one might think it was hyperbolic, or tinged with sci-fi dystopian.

On May 24, 2020, a black man named George Floyd died while being arrested by a white police officer, who held him down with a knee to the back of his neck. The scene was captured on cell phone videos and posted on social media, where it went viral. What followed was a cascade of protests in 140 cites across the country, which many turned into violent riots, looting and burning buildings. The explosive reaction wasn't just about that isolated incident. The George Floyd killing was simply the catalyst for an underlying social anger that boiled over across the country. Race is but one of many flashpoints in America's sociopolitical dynamic. Income disparities are now so pronounced that America's richest 1 percent of households averaged 139 times as much income as the bottom 20 percent.[11] We also have about 400 million privately owned firearms in circulation.[12] Add to these multiple incendiary factors, tens of millions of Americans who had their livelihoods replaced by AI, and the amount and level of uncontrolled violence is surreal.

Bread and Circus: The Rise of the Useless Class

In the not-so-distant future, as AI-fueled automation renders human labor increasingly obsolete, the U.S. government rolls out an unprecedented Universal Basic Income (UBI) program. Hailed as a solution for the millions of displaced workers, it's promoted as a compassionate response to the economic dislocation that has left the middle and working classes jobless, redundant, and lost. The reality is anything but humane. UBI, alongside a steady stream of mind-numbing entertainment, forms the backbone of a sinister new social order—a "bread and circus" system designed to placate the masses while stripping away any remaining sense of purpose, agency, or ambition.

Behind this facade, a seismic shift takes place. While the average American is pacified with just enough income to survive and access to immersive virtual reality centers in every home, a small group of hyper-wealthy elites fortify their own future—secure, secluded, and endlessly advantaged. These elites, concentrated in Silicon Valley, exist in fortified enclaves. They embrace the latest in neurotechnology, using devices like Elon Musk's neural chips to enhance their mental and physical capabilities, effectively transcending human limitations and edging toward a state of "cyborgism." To them, the potential of artificial intelligence and advanced biotechnology means unparalleled productivity, wealth, and lifespan. But to the broader population, it means exclusion, obsolescence, and a life devoid of hope or purpose.

As the wealth gap morphs into an unbridgeable chasm, traditional institutions fall apart. The military, too, evolves into an elite, autonomous faction fueled by AI-driven weaponry and robotic combat units, its loyalty resting primarily with the techno-oligarchs funding its operations. National defense is reoriented toward an ongoing cold war with China, which has also embraced AI in its race for dominance. Similarly,the political system, is reduced to a puppet show, a mere shadow of governance, with decisions and policies increasingly orchestrated by a coalition of tech moguls whose allegiances lay with innovation and profit, rather than the democratic ideals they purported to uphold.

In this engineered landscape, America's once-thriving heartland has become a wasteland of human apathy and lost potential. With UBI ensuring basic survival, people find themselves in a paradox of stagnation. Their essential needs are met, but the drive, the spirit, the hope that once defined American society has been systematically stripped away. The VR headsets in their homes become modern-day shackles, a virtual escape that only deepens the disconnect from reality, replacing authentic human experience with simulated fulfillment.

Now, America's vast "useless class" drifts through days without purpose, a mass of disillusioned individuals anesthetized by technology. For the first time in history, productivity is reaching unimaginable heights, yet nearly all of its rewards are concentrated within the hands of the top 0.1%. The elites have found a way to ascend, embracing transhumanism and augmenting themselves, while the masses—deemed irrelevant—languish in state-sponsored complacency. The soaring advancements of a few contrast sharply with the passive resignation of the many, creating a chasm not only of wealth but of humanity itself.

As the elites continue to push the boundaries of human potential and artificial intelligence, the great majority of Americans are left to wither in an era of technological abundance that has somehow bypassed them entirely. Bread and circus keeps them quiet, but their silence speaks volumes about a society that has abandoned them, and in the process, has lost its very soul.

I'm going to close this chapter with a few words as they say, straight from the horse's mouth, former OpenAI engineer Leopold Aschenbrenner: "Everyone is now talking about AI, but few have the faintest glimmer of what is about to hit them. Mainstream pundits are stuck on the willful blindness of 'it's just predicting the next word'. They see only hype and business-as-usual; at most they entertain another internet-scale technological change. Before long, the world will wake up. But right now, there are perhaps a few hundred people, most of them in San Francisco and the AI labs, that have situational awareness. Through whatever peculiar forces of fate, I have found myself amongst them. A few years ago, these people were derided as crazy—but they trusted the trendlines, which allowed them to correctly predict the AI advances of the past few years. Whether these people are also right about the next few years remains to be seen. But these are very smart people—the smartest people I have ever met—and they are the ones building this technology. Perhaps they will be an odd footnote in history, or perhaps they will go down in history like Szilard and Oppenheimer and Teller.

If they are seeing the future even close to correctly, we are in for a wild ride."

I'll decode that pleasantry for you.

The "smartest people" Aschenbrenner is referring to is the elite club he belongs to. No doubt, they are smart. And as he also points out, "they are the ones building this technology." He is gleeful about "the wild ride we are in for." But see, the ride will be quite different for him and the other "smartest people" then it will be for us. They're driving the AI bus, and we're just passengers at the back. As Geoffrey Hinton points out, "I don't see how this turns out well for humanity." He's also one of the "smartest people," so the humanity he's referring to is *us*, the ones at the back of the bus with no control of where it's taking us.

It's time to take control.

Chapter IV.

Fortune Favors a Prepared Mind

In a world on the brink of widespread AI transformation, preparation is not a luxury—it's a necessity. This isn't about alarmism or conspiracy; it's a reality that requires a proactive approach to stay grounded, resilient, and ready for the radical social change ahead.

One fundamental practice for building mental strength and resilience is establishing a morning routine.

Studies show[13] that a consistent morning routine can reduce stress, boost energy, and enhance mental preparedness. A good morning routine doesn't need to be complex; it's about creating a short, repeatable structure that begins the moment you wake up. Experiment to find what suits you—whether it's meditation, deep breathing, a cold-water bath, or something as simple as a walk. The goal is to create a routine that centers you, providing a space away from the noise and opinions that constantly bombard us. Make it personal. Make it powerful. Cultivate your own internal strength.

Health as a Form of Resistance

In a time when machines are on the verge of reshaping humanity, nothing is more revolutionary than maintaining your health and vitality. With three-quarters of Americans now overweight or obese,[14] and half of our population with diabetes or pre-diabetes,[15] we live in a society that has been engineered to profit from our poor health. For decades, corporate food chains and Big Pharma have treated the public like cash cows, promoting unhealthy diets and then promoting expensive drugs for the diseases they've caused.

A Big Mac, French fries, and a Coke is 1,360 calories, 58 grams of saturated fat and 95 grams of sugar.[16] In essence, it is diabetes, cancer, and heart disease slickly packaged by profiteers.

For 50 years, we have been living in a supercycle of self-inflicted poor health outcomes. It is a death spiral unless we break free. It's simple. When you are unhealthy you become an indentured servant to the medical industrial complex, a massive machine with a profit-over-patient ethos.

It is important to find your own path to health. But there are three simple pillars of health that can fortify us against this machine-driven era: diet, exercise, and purpose.

1. Diet: Let's stick to the facts—chronic inflammation is a root cause of diseases like diabetes and cancer,[17] and a plant-based diet is proven to reduce it. Aim for at least a 90% plant-based diet to lower inflammation and strengthen your body. Fast food is poison; the corporations want you addicted, but you can choose to break free. It's no accident that the elites building AI systems are thin, often plant-based eaters. Choose what they choose, not what they sell.

2. Exercise: Make movement a daily priority. A 30-minute routine that includes cardio and strength training is enough to keep your body resilient. Find a rhythm that works for you and stick to it. The system wants you sedentary, dependent on unnecessary medications—but you owe it to yourself to stay active and independent.

3. Purpose: This might be the most important pillar of all. The longest-living communities in the world cite purpose as a key ingredient to a fulfilling life. Find something that ignites you—a

hobby, a skill, a cause. Purpose is your anchor, especially as technology threatens to strip us of what makes us human.

This isn't easy, but it's simple. Diet, exercise, and purpose—three basic practices that empower you to resist the system, defy the machines, and maintain your humanity. This is a new kind of class warfare, one that demands not just awareness but action. Embrace these principles, and strengthen yourself for what lies ahead.

Prepare for the Inevitable

The confluence of AI-driven job displacement and a $35 trillion national debt signals an unprecedented economic challenge. Our fiat currency system, which enables unlimited money printing at the whim of politicians, has created a fiscal vortex that threatens long-term financial stability. In this environment, Bitcoin emerges as a compelling alternative—a mathematical fortress immune to political manipulation. Its core attributes—fixed supply, decentralized control, and trustless verification—represent a deliberate shift of financial power from institutions to individuals. This isn't just theory anymore: the entry of financial giants like BlackRock and Vanguard into the Bitcoin market signals a seismic shift in institutional acceptance. Bitcoin's foundation in immutable mathematics, rather than human discretion, offers a rare certainty in increasingly uncertain times.

Unplug From the Matrix

Let me briefly describe Bitcoin's revolutionary journey and significance in the financial world:

Bitcoin emerged in 2009 when Satoshi Nakamoto published the whitepaper "Bitcoin: A Peer-to-Peer Electronic Cash System," introducing the first decentralized cryptocurrency. The timing was significant, coming right after the 2008 financial crisis when trust in traditional banking systems was at a low point.

Bitcoin's fundamental innovation is its ability to enable trustless, peer-to-peer transactions without intermediaries through its proof-of-work consensus mechanism. Miners compete to solve complex mathematical problems, securing the network and validating transactions. This process creates new bitcoins and maintains the blockchain's integrity.

The cryptographic foundation of Bitcoin is remarkably robust. It uses SHA-256 hashing and public-key cryptography, making it practically impossible to forge transactions or alter the blockchain's history. Each block builds upon previous blocks, creating an immutable chain of transactions.

Bitcoin's limited supply of 21 million coins makes it a compelling store of value, often called "digital gold." Unlike fiat currencies that can be printed at will, Bitcoin's scarcity is mathematically guaranteed. This fixed supply, combined with growing demand, has driven extraordinary price appreciation--from essentially zero in 2009 to tens of thousands of dollars per coin today.

The institutional adoption of Bitcoin has been transformative. BlackRock's spot Bitcoin ETF marks a pivotal moment in Bitcoin's journey from a fringe technology to a mainstream financial asset. Major corporations are adding Bitcoin to their balance sheets, and traditional financial institutions are increasingly offering Bitcoin services to their clients.

The network has proven its resilience through multiple market cycles, growing stronger with each test. Its decentralized nature means no single entity can control or shut down the network, providing a level of financial sovereignty previously impossible.

Looking ahead, Bitcoin's role as a hedge against inflation and monetary policy uncertainty positions it as a crucial part of modern investment portfolios. With growing institutional acceptance and improving infrastructure, Bitcoin appears poised for continued growth and adoption.

The potential for Bitcoin and blockchain technology to revolutionize international trade is immense. Cross-border transactions that traditionally took days and incurred high fees can now be settled in minutes at minimal cost. This efficiency could fundamentally transform global commerce, reducing friction in international business operations.

Bitcoin's primary strength remains as a digital store of value and medium of exchange. Its growing acceptance by federal authorities, combined with institutional adoption, suggests we may be entering a new era where Bitcoin plays a central role in both personal and national financial strategies.

The potential integration of Bitcoin into traditional financial systems could lead to more efficient markets, reduced transaction costs, and greater financial inclusion globally. As regulatory clarity improves, we will see even more innovative applications of Bitcoin in international trade and finance.

As you can see, I'm bullish on Bitcoin for multiple reasons, primarily because I find its philosophy and structure to be perfectly suited for the uncertainties of the future. Whatever financial vehicle(s) you choose as a hedge against the impending economic crisis, think it through carefully. But act before it's too late.

Preparing for Financial Turbulence: Your Survival Guide

True financial resilience isn't built overnight – it's a strategic fortress constructed with purpose and foresight. Here's your essential action plan:

1. Master Your Liquidity: Keep enough cash readily available to weather 6-12 months of

disruption. Think of this as your financial oxygen tank. You can't swim deep without it.

2. Secure Your Value: Diversify into hard assets like Bitcoin and gold that historically shine during currency crises. These are your financial anchors in stormy seas.

3. Slay Your Debt: Aggressively eliminate high-interest obligations. Each debt is a leak in your financial boat. Plug them before the waters get rough.

4. Build Your Brain Trust: Partner with a forward-thinking financial advisor who understands both traditional and digital assets. Regular strategy sessions are crucial.

5. Create Your Circle: Develop a network of financially savvy friends and family. These connections are your economic early warning system and support structure.

6. Secure Your Communications: Establish encrypted channels (like ProtonMail) with your trusted circle. Financial privacy isn't paranoia, it's prudence.

7. Draft Your Battle Plan: Document your critical expenses and create multiple contingency

budgets. Know exactly how to scale down if needed.

Think of your financial health like a centuries-old redwood tree: it needs deep, strong roots (your fundamentals), a solid core (your store of value), and multiple branches (your diverse strategies) to survive any storm. The time to plant and strengthen that tree isn't during the hurricane, it's now, while the skies are still clear enough to see the horizon.

Call to Action

Time is running out. A friend of mine who has two small children was talking about saving for their college tuitions. His kids won't be of college age for at least 10 years. When I said colleges won't exist by then, degrees will be useless, at least in the form they are today, his face went blank. "What will they do...my children?" he asked.

People are in dangerous denial about the rapid social transition that's occurring in plain sight. Repetitive motion jobs, communication, transportation, warehousing, fast-food, retail, gone for the most part. Cognitive skill jobs, white collar jobs in finance and tech will be nonexistent for the most part. Some of the touchy-feely jobs in healthcare will still be safe by then. Maybe. Right now, a doctor friend uses ChatGPT (and AGI is right around the corner!) for most of his diagnostic work and drug interaction information. Radiologic scanning and pathology will all be AI driven. So even the once "safe" healthcare market will be drastically reduced.[18]

We are in a nationwide delusion that reminds me of the famous scene from the film, *A Few Good Men*, when the lawyer (Tom Cruise) who was cross-examining the Colonel (Jack Nicolson) says "I want the truth!" to which the Colonel shouts, "You can't handle the truth!"

We *need* to handle the truth. And we need to harden our minds to it.

AI: Not My Master

The coming wave of AI displacement is inevitable—but how it affects us is a choice. Our choice. But only if we unite, organize, and act now. Here's how we build this movement:

Digital Mobilization:

- Flood social platforms with #AINotMyMaster

- Share personal stories of AI impact on work and community

- Create viral content showing the real human cost of unchecked automation

- Host weekly Twitter Spaces and Reddit AMAs on AI accountability

Grassroots Power:

- Form local action groups in every community

- Share this handbook neighbor-to-neighbor, friend-to-friend

- Organize town halls and community forums

- Build coalitions with labor unions and workers' rights groups

Visible Resistance:

- Tout the message "AI: Not My Master"

- Put up posters in local businesses and community centers

- Create eye-catching public art installations

- Organize coordinated national days of awareness

Media Amplification:

- Launch a network of podcasts focused on human-first automation

- Get local radio stations to host weekly discussions

- Create compelling video content for YouTube and TikTok

- Partner with sympathetic journalists and publications

Mass Mobilization:

- Plan a Million Worker March on Washington

- Coordinate simultaneous demonstrations in major cities

- Stage peaceful protests at tech company headquarters

- Create a visible presence at all political town halls

Remember: Every great movement in history started with ordinary people who refused to accept the unacceptable. The Civil Rights Movement didn't just happen--it was built person by person, day by day, by people exactly like you and me.

This isn't just about jobs. This is about human dignity. About who we are and what we'll become. About whether technology serves humanity, or the other way around.

The window for action is closing.

Every day we wait is another day the AI agenda advances unchallenged. But together, we can ensure that the future of work has a human heart.

Stand up. Speak out. Act now.

Because if not us, who? If not now, when?

AI: Not My Master. The choice is ours.

Conclusion

Humans are mere spectators to an AI narrative being written in the shadowed corridors of Silicon Valley—a story we've been excluded from, but one that fundamentally transforms our world. We demand transparency. We demand the full story, in real time.

Just as Bushido was the sacred code that bound the samurai and defined Japanese society, we now require our own sacred covenant—a unified ethical framework to navigate this technological revolution.

America was built on utility, forged by the most productive and innovative workforce in human history. We are the descendants of hard pioneers, of builders and disruptors. We cannot—we will not—surrender our agency to 500 cloistered engineers and their algorithmic machines. The spirits of those who carved civilization from wilderness would not accept passive obsolescence.

Our resistance is not measured by certain victory, but by the courage to stand against what seems inevitable. It is better to fight against something you might not beat, than to fade silently into irrelevance.

AI: Not my Master. Not my *fucking* master!

References

[1] https://www.nytimes.com/2023/05/30/technology/ai-threat-warning.html

[2] https://situational-awareness.ai/

[3] https://www.ibm.com/think/topics/technological-singularity

[4] https://www.amazon.com/Artilect-War-Controversy-Concerning-Intelligent/dp/0882801546

[5] https://ourworldindata.org/moores-law

[6] https://en.wikipedia.org/wiki/Moore%27s_law

[7] https://ideas.ted.com/how-the-gains-we-make-in-ai-could-ultimately-destroy-us/

[8] https://situational-awareness.ai/

[9] https://shorturl.at/Nfj8S

[10] https://healthcostinstitute.org/hcci-originals-dropdown/all-hcci-reports/kaiser-health-news-1-3

[11] https://inequality.org/facts/income-inequality/

[12] https://www.thetrace.org/2023/03/guns-america-data-atf-total/

[13] https://pmc.ncbi.nlm.nih.gov/articles/PMC9529170/

[14] https://www.thelancet.com/journals/lancet/article/PIIS0140-6736(24)01548-4/fulltext

[15] https://www.aicr.org/news/half-of-us-adults-have-diabetes-or-prediabetes/

[16] https://shorturl.at/xFjUs

[17] https://www.nature.com/articles/s41591-019-0675-0

[18] hhttps://shorturl.at/QAyRwat/QAyRw

Made in the USA
Monee, IL
04 November 2026

67377414R00036